A Handbook Of Tricuspid and Pulmonary Valve Disease

Alok Ranjan
MD, DNB, MRCP (UK), DM (Card.)
Sr. Consultant - Cardiology
Wockhardt Hospitals
India

authorHOUSE®

AuthorHouse™
1663 Liberty Drive
Bloomington, IN 47403
www.authorhouse.com
Phone: 1-800-839-8640

© 2012 Dr. Alok Ranjan. All rights reserved.

No part of this book may be reproduced, stored in a retrieval system, or transmitted by any means without the written permission of the author.

Published by AuthorHouse 12/11/2012

ISBN: 978-1-4772-9057-6 (sc)
ISBN: 978-1-4772-9056-9 (hc)
ISBN: 978-1-4772-9055-2 (e)

Library of Congress Control Number: 2012921604

Any people depicted in stock imagery provided by Thinkstock are models, and such images are being used for illustrative purposes only.
Certain stock imagery © Thinkstock.

This book is printed on acid-free paper.

Because of the dynamic nature of the Internet, any web addresses or links contained in this book may have changed since publication and may no longer be valid. The views expressed in this work are solely those of the author and do not necessarily reflect the views of the publisher, and the publisher hereby disclaims any responsibility for them.

Disclaimer

Medicine is a constantly changing science. New research findings necessitate continual changes in disease concept and its management. The author and publisher of this handbook have used reasonable efforts to provide up-to-date, accurate information that is within generally accepted medical standards at the time of publication. However, as medical science is ever evolving, and human error is always possible, the author and publisher (or any other involved parties) do not guarantee total accuracy or comprehensiveness of the information in this handbook, and they are not responsible for omissions, errors, or the results of using this information. The reader should confirm the accuracy of the information in this handbook from other sources. In particular, all drug doses, indications, and contraindications should be confirmed in package inserts.

The author has made every effort to trace the copyright holders for borrowed material. If he has inadvertently overlooked any, he will be pleased to make necessary arrangement at the first opportunity.

Dedicated to my teachers and friends in
premedical school and colleges

Those who educate children well are more to be honored
than parents, for these only gave life,
those the art of living well.
'Aristotle'

Contents

The Tricuspid Valve Complex xi

Tricuspid Stenosis (TS) 1
Etiology 3
Hemodynamics 5
Clinical features 7
Investigations 11
Management 15

Tricuspid Regurgitation (TR) 17
Etiology 19
Clinical Features 21
Investigations 23
Management 27

Ebstein's Anomaly of Tricuspid Valve 31
Background 33
Pathophysiology 39
Clinical features of Ebstein's anomaly 41
Natural History 55
Management 59
Pulmonary Valve 65

Pulmonary Stenosis (PS) 67
Etiology 69
Clinical Features 71
Investigations 75
Management 81

Pulmonary Regurgitation (PR) 85
Etiology 87
Clinical Features 89
Investigations 91
Management 93

Straight Back Syndrome 95

Suggested Reading 99

Abbreviations 101

About the Author 107

The Tricuspid Valve Complex

The tricuspid valve complex consists of three leaflets, the chordae tendinae, two discrete papillary muscles, the fibrous tricuspid annulus, and the right atrial and right ventricular myocardium.

Leaflets: There are three sail-like leaflets: anterior, posterior, and septal. All are attached to the tricuspid valve annulus. The anterior leaflet is the largest and is the most anatomically constant of the three, whereas the posterior leaflet is notable for the presence of multiple scallops. The septal leaflet is the smallest and arises medially directly from the tricuspid annulus above the interventricular septum. Because the small septal wall leaflet is fairly fixed, there is little room for movement if the free wall of right ventricular/tricuspid annulus should dilate. Dilation of the tricuspid annulus therefore occurs primarily in its anterior/posterior (mural) aspect, which can result in significant functional TR as a result of leaflet malcoaptation. The septal aspect of the tricuspid annulus is considered to be analogous to the intertrigonal portion of the mitral annulus in that it is relatively spared from annular dilation. Because of this property, tricuspid annular sizing algorithms have been based on the dimension of the base of the septal leaflet

The normal tricuspid valve thickness is less than or equal to 3 mm

Papillary muscles and chordate tendinae: The anterior papillary muscle provides chordae to the anterior and posterior leaflets, and the medial papillary muscle provides chordae to the posterior and septal leaflets. The septal wall gives chordae to the anterior and septal leaflets (there is no formal septal papillary muscle). In addition, there may be accessory chordal attachments to the right ventricular free wall and to the moderator band, a feature distinguishing the right and left ventricles. These multiple chordal attachments are important mediators of TR, as they impair proper leaflet coaptation in the setting of right ventricular dysfunction and dilation.

Annulus: The tricuspid annulus has a complex 3-dimensional structure, which differs from the more symmetric "saddle-shaped" mitral annulus. This distinct shape has implications for the design and application of currently available annuloplasty rings in the tricuspid position (most currently available rings are essentially planar). The tricuspid annulus is very dynamic and can change markedly with loading conditions. Even during the cardiac cycle, there is a $\approx$19% reduction in annular circumference ($\approx$30% reduction in annular area) with atrial systole

Tricuspid Stenosis (TS)

Etiology

Rheumatic Heart disease

    Rheumatic TS is uncommon and almost never occurs as an isolated lesion. Hence, TS in a case of RHD is indicative of more extensive disease

    It is also almost never seen with isolated MR or isolated aortic valve disease. *MS almost always coexists with TS.*

    Organic TR is always associated with TS

        Causes of TS with TR
            RHD
            Carcinoid disease
            Ebstein's anomaly

Congenital
Carcinoid disease
Fibroelastosis
EMF
SLE (Systemic lupus erythematosus)

Rheumatic TS

Incidence: Incidence of TS in a case of RHD

    Autopsy series: 22 – 44 %
    Clinical series: 14 %;

Hence TS is either likely to be missed clinically in many patients or is not severe enough to produce clinical features.

Studies:
  Kitchin and Turner: 3.1 %
  Kinare et al: N= 150 (autopsy study)
    Tricuspid valvulitis: 34 %
    Significant TS: 3.9 %
  Roy and Tandon:
    Tricuspid valvulitis: 42.4 %.

For TS to be diagnosed, TV circumference should be less than 10 -11 cm and the area should be less than 7 sq. cm. For clinical recognition TVA should be less than 4.9 sq. cm and TV circumference should be less than 8 cm.

(Kitchin & Turner).

Hemodynamics

If both MS and TS coexist and are of equal severity, TS is the limiting factor in cardiac output.

Variation of PVR with exercise:

> In normal persons, PVR decreases with exercise
> In pure MS: PVR increases with exercise
> In combined MS with TS: Usually decreases but mixed response may be seen.

Clinical features

*Eliciting symptoms and signs which point to **'atypical severe MS'** is often a clue to coexisting TS.*

> Elevated systemic venous pressure and symptoms due to TS
> > Abdominal swelling (ascites),
> > Edema and anasarca,
> > Abdominal pain (due to hepatomegaly) and
> > Fatigue (decreased cardiac output)
>
> All these symptoms are out of proportion to dyspnea
> > (which the most prominent symptom of MS).
>
> *The absence of symptoms and signs of pulmonary congestion in a patient with obvious MS should **suggest** possibility of associated TS
>
> *In presence of TS, signs of MS are masked and it should be taken as clue to presence of TS.

Pulse

> Sinus rhythm (**surprisingly common in patients with TS**)

JVP

Flicking A wave
- Pathognomonic of TS
    - Best seen in External jugular vein
    - Can be palpated also
    - It is either presystolic or almost synchronous with S1 (Unlike systolic S wave of TR which follows carotid pulsations)
    - If 2 distinct waves are seen at 45° and A wave is more than V wave and the mean JVP is high then *severe* TS is unlikely and in most cases dominant TR is present.
- If atrial fibrillation is present, diagnosis of TS is not possible from JVP.

Auscultation

Murmur
- Mainly presystolic, but usually lacks the crescendo quality of MS (as T1 is not as loud as M1)
- Mid diastolic murmur: usually short (as compared to MS, because, TTG is not as high as TMG)
- Intensity of murmur increases with inspiration (ask patient to take long, slow, deep breath and hold)
- Duration of MDM is not proportional to severity of TS (compare from MS) but very short, insignificant MDM is not associated with significant TS
- Best heard at lower sternal border a little to the left of midline but never maximal at tricuspid area (TV is usually displaced by large RA and rotation of heart)

Rarely in severe combined TV disease (TS with TR), pansystolic murmur of TR may decrease with inspiration and increase with expiration.

Other sign: Presystolic hepatic pulsation

Important signs that should raise suspicion of TS in a case of RHD:

1. Tall 'a' wave in JVP with slow and shallow 'y' descent

2. Absence of signs of pulmonary hypertension in an obvious cases of significant MS

    Inconspicuous left parasternal lift
    P2: Not palpable
    Clear lung fields

3. Plethora of signs of systemic venous congestion and paucity of signs of pulmonary venous congestion

    Presence of ascites and anasarca without lung signs

4. Presystolic hepatic pulsation

Investigations

1. ECG

Tall peaked P wave
> **Clue for coexisting TS:**
> RAE disproportionate to degree of RVH

Tall 'p' wave can be due to TS or PH (in case of associated MV disease).
> CR I lead is superior to conventional leads in diagnosing RAE (> 2 mm amplitude is taken as criteria for RAE)

RAE in a case of MS
> Can be present without evidence of TS in 50 % cases; but commonly associated with signs of RVH on ECG
> If P wave height is > 2.5 mm: TV disease (Associated organic TS and / or TR is very likely)
> If > 3.0 mm: always signifies a significant organic TV disease.

Prolonged PR interval;
> Due to RA dilation

QRS amplitude:
> Reduced amplitude of QRS in lead V1 (Often with q

wave) with tall QRS in V2.

Due to presence of massive RAE, ventricular septum is rotated and presence of large RA between IVS and lead, affects amplitude.

2. CXR:

RAE: Right heart border > 5.0 cm **from midline with relative or absolute lack of pulmonary artery shadow** and relatively clear lung fields are suggestive of TS.
RAE alone is unreliable indicator of TS on CXR
Calcification of TV is **not** seen (Calcification of TV is surprisingly **rare** as compared to MV calcification)

3. Echocardiography:

M Mode features:
    Findings similar to MS
        Reduced EF slope
        Thickening and tethering of leaflets

2 D Features:
    Doming of valve: Hallmark of TS
    Restricted motion and thickening of leaflets (also seen in other condition e.g., Carcinoid syndrome)
    TVA can not be measured (by planimetry) as all 3 leaflets of TV can not be focused in any single view (unlike MV)

2 D features are 100 % sensitive and 90 % specific for diagnosis of TS

Color Doppler features
    Normal Tricuspid inflow velocity
        Peak: 0.3 – 0.7 m/s (< than 1 m/s always)
        Mean gradient: < 2.0 mm Hg
    An end diastolic gradient of > 2 mm Hg indicates TS
        > 7 mm Hg indicates severe TS
    TVA by PHT method

Can be used
Some recommend 190 in place of 220 for calculation (220 used for MS) 190 / TV PHT

4. Cardiac catheterization:

Intra Cardiac pressures:
If a trans tricuspid gradient (TTG) exists during diastole: TV (either TS and / or TR) disease is present.
TTG should be measured with simultaneous recording with 2 catheters; one each in RA and in RV.

Killip and Lucas criteria of TS:
If mean TTG is more than 1.9 mm Hg at rest, it is significant and indicates TS.
Magnitude of TTG bears no relation with severity of TS as associated TR may also cause increased TTG
Usually TTG > 5 mm Hg is suggestive of significant TS (but may not be true in presence of TR).

RA wave forms:

Organic TR (TR associated with TS)
Normal X descent is retained
Y descent is slow
Systolic (S) wave is late and less conspicuous
Functional TR
X descent is reduced or obliterated and replaced by systolic (S) wave
Steep Y descent
In sinus rhythm:
If A wave is more than S wave by 5 mm Hg: Suggestive of severe TS
If S wave is more than mean RAP then severe TR is likely
In atrial fibrillation:
Wave form is less helpful

Most reliable evidence of TV obstruction comes from the changes of pressure with respiration

Valve area calculation:
  By Gorlin formula
    Not so accurate

Management

Indication for Intervention:

    Tricuspid valve gradient (mean) > 5 mm Hg
    Tricuspid valve area < 2 sq. cm

Options:

1. Balloon tricuspid valvotomy

2. Surgery:

    OTV (Open Tricuspid Valvotomy): Tricuspid valve is converted to functionally bicuspid valve by opening the 2 out of 3 commissures; the commissure between ATL and PTL is not opened to prevent development of severe TR post operatively.
    Tricuspid Valve Replacement: Bioprosthetic valve **only**. Mechanical valves have very high incidence of valve thrombosis at tricuspid and pulmonary position; surprisingly tilting discs have higher incidence of thrombosis as compared to caged – ball valve.

Tricuspid Regurgitation (TR)

Etiology

Rheumatic Heart disease
  With pulmonary hypertension (Functional TR)
  In absence of PH (Organic TR)
    Systolic pulmonary artery pressures greater than 55 mm Hg are likely to cause TR with anatomically normal tricuspid valves, whereas TR occurring with systolic pulmonary artery pressures less than 40 mm Hg is likely to reflect a structural abnormality of the valve apparatus.
Pulmonary hypertension from any cause
Trauma
EMF (Endomyocardial fibrosis)
Ebstein's anomaly
Congenital isolated tricuspid insufficiency
Tricuspid valve prolapse
Endocarditis
Ischemic heart disease
Tumor
Rare causes:
  Pacemaker wires
  Hyperthyroidism
  Loffler's endocarditis
  Sinus of Valsalva aneurysm

Clinical Features

Symptoms:

>Predominantly due to other coexisting valvular lesions
>May complain of progressive fatigue, edema and anorexia
>Lack of **paroxysmal** pulmonary symptoms
>**Wasting**: causes
>>Anorexia
>>Hyper metabolic state
>>Malabsorption
>>Protein losing enteropathy
>>Immunodeficiency state

Signs:

Jugular Venous Pulse in Tricuspid Regurgitation

Feature	Findings	Mechanism / significance
Level	Normal or elevated	Elevated with RV failure
		Associated organic TS

Waves		
A wave	Prominent	Prominent A wave with Severe PAH
		Associated organic TS
		May merge with 'v' wave
X descent	Preserved	Mild TR or Larger RA
	Obliterated	Obliterated with Severe TR
		Smaller RA
V wave	Very prominent or Normal	Prominent V wave with Severe TR
		Smaller RA
Y descent	Steep and Rapid	Rapid filling of RV in TR
	Slow	Associated TS
H wave	Can be prominent	Occurs after rapid filling wave of RV and (mistaken for 'a' wave) may be reflected in RA

Murmur:

    PSM
        Best heard at lower sternal border
        Inspiratory accentuation (Carvallo's sign)
            Carvallo's sign is absent if
                CHF or right heart failure is present
                Venous pressure is very high
                Severe TS is also present
                AF

    Systolic hepatic pulsation
    Peripheral edema
    Peripheral cyanosis

Investigations

1. **ECG** :

    IRBBB
    'q' in V1: Related to RAE;
    AF: May be present in 70 % cases
    In multivalvular disease, it is rare to find evidence of RAE on ECG

2. **CXR**:

    Cardiomegaly
    Disproportionate enlargement of right sided cardiac chambers including RA
    Relatively clear lung fields
    Signs of other concomitant valve lesions

3. **Echocardiography**

    Organic TR
    Structural disease of tricuspid leaflet and / or subvalvular apparatus.
    Functional TR
    Secondary to pulmonary hypertension
    No structural deformity of TV
    If annulus is greater than 4 cm or TR is severe
    Importance: Annuloplasty should be advised

M Mode features: Non specific
- Volume overload of RV and RA
- Paradoxical motion of the ventricular septum is observed and is similar to that found in an atrial septal defect

2 D features
- Structural deformity of TV
- Volume overload of RV and RA
- Paradoxical motion of the ventricular septum is observed and is similar to that found in an atrial septal defect

Echo criteria of Severe TR
- Systolic RA annulus diameter > 30 mm
- Diastolic RA annulus diameter > 21 mm / m2 and less than 25 % degree of systolic annular shortening

Pulse wave Doppler
- A jet > 30 mm from annulus
- TR area > 4 cm2

Color Doppler
- Regurgitant jet area / RA area
  - Severe TR: Jet area > 33 % of RA
  - Moderate TR: Jet area 20 - 33 % of RA
  - Mild TR: Jet area less than 20 % of RA

IVC flow pattern on echocardiography

Normal flow pattern in IVC:
- A predominant negative* systolic component (s),
- Negative diastolic component (d) of smaller amplitude than the 's' component,
- Ends with onset of the atrial contraction which may produce a brief flow reversal, small positive wave.
- In the patient with atrial fibrillation the relative amplitude of the systolic component (s) is smaller and the atrial component disappears.

* Blood flow approaching the transducer is marked as positive and flow running away as negative. The curves are best recorded during expiratory apnea.

IVC flow pattern in tricuspid regurgitation-

Mild TR:
The amplitude of the systolic component (s) is decreased
Moderate or severe TR:
The normal systolic anterograde component (s) is replaced by a positive systolic retrograde component.

Flow patterns in hepatic veins:

Normal flow:
During most the cardiac cycle, flow in the hepatic veins of normal subjects should be in the anterograde direction (produces a negative wave). During systole and diastole, a drop in pressure in the right atrium leads to two anterograde waves (similar to two negative waves of 's' and 'd' in IVC flow pattern).

Pattern in tricuspid regurgitation:
Depending on the degree of incompetence of this valve, the systolic flow in the hepatic veins may be either decreased or reversed
Due to the incompetence of the tricuspid valve, the high right ventricular systolic pressure is transmitted to the right atrium and hepatic veins.

Sakai et al. have divided hepatic vein Doppler waveforms in tricuspid regurgitation into three types according to systolic flow wave changes:
Type 1;   The amplitude of negative component is smaller than the normal amplitude
Type 2:   There is no systolic flow; and

Type 3: Systolic flow is reversed i.e., a positive wave is recorded.

They found good correlation between moderate and severe tricuspid regurgitation and type 3 and between types 1 and 2 and mild or absent tricuspid regurgitation.

4. Cardiac Catheterization

Mean RAP > 10 mm Hg
Ventriculography shows grade 3+/4 TR
    Systolic annular diameter > 27 – 35 mm

5. Digital examination during operation: Less accurate

Regurgitant jet only 1-2 cm from valve: Mild TR
Regurgitant jet more than 3 cm from valve: Moderate TR

Management

Functional TR

> Characterized by a dilated annulus with normal leaflets and chordal structures

> The decision to correct functional TR depends on the severity of TR and level of PVR (pulmonary vascular resistance)

Severity of TR and PVR	Treatment
Mild TR and PVR < 500 dyne-sec/cm5	Left alone; usually resolves after left sided valve correction
Moderate TR with low PVR	Left alone; usually improves after surgery of left sided valve; Some argue that patient will do clinically better if TR is repaired regardless of PVR
Severe TR	Should always be repaired

Recommendation for surgical correction of tricuspid regurgitation (TR) Indication Class

1. Tricuspid repair or replacement for severe primary or sec-

ondary TR, in symptomatic patients not responding to medical treatment (I B)

2. Tricuspid repair or replacement for severe TR in patients requiring mitral valve surgery, particularly in the presence of pulmonary hypertension (mean pulmonary artery pressure >50 mmHg) or right ventricular dilation and dysfunction (I B)

3. Tricuspid repair for moderate functional TR, secondary to left-heart lesion at the time of mitral valve surgery (IIa C)

Contraindication

4. Isolated valve replacement or repair for TR, in an asymptomatic patient with normal right ventricular function (III C)

Surgical approaches

> Rigid or flexible annular bands (open or closed),
> Reduce annular size and achieve leaflet coaptation,
>> The open ring shown spares the atrioventricular node (AVN), thus reducing the incidence of heart block
>> Ring annuloplasty
>>> Carpentier-Edward (rigid)
>>> Duran ring (Flexible)
>>>> Many believe that implantation of a ring is specifically indicated in organic disease
>>>> In rheumatic heart disease (TR). due to associated TS, commissurotomy should be combined with ring annuloplasty.

> Posterior annular bicuspidalization.
>> Less commonly used
>> This surgical technique places a pledget-supported mattress suture from the anteroposterior commissure to the posteroseptal commissure along the posterior

annulus. This is based on prior studies by Deloche et al that show posterior annulus dilation occurs in functional TR and that a focal posterior tricuspid annuloplasty can be effective in selected cases.[5]

Other approaches
    Edge-to-edge (Alfieri-type) repairs
    Partial purse-string suture techniques to reduce the anterior and posterior portions of the annulus (DeVega-style techniques). DeVega and flexible band annuloplasties appear to have a lower freedom from recurrent TR than rigid annuloplasty rings.
    De Vega annuloplasty
    Semicircular annuloplasty
    Does not involve foreign material (suture annuloplasty)
    Leaves a pliable annulus
    It aims at reducing the size of the annulus in the segments that are prone to dilatation and those corresponding to the anterior and posterior leaflets.
    Rapid but may not be as durable as prosthetic rings

<u>Rigid annuloplasty should be the preferred surgical approach for significant TR if the leaflets are spared from the disease process.</u>

Mortality

Isolated TVR	5 %
MV surgery with TVR	10 %
AVR with MVR with TVR	20 %

    High mortality in combined surgery is due to advanced RV dysfunction

Ebstein's Anomaly of Tricuspid Valve

Background

Introduction:

Characterized by a downward displacement of the tricuspid valve into the right ventricle, due to anomalous attachment of the tricuspid leaflets, the Ebstein tricuspid valve tissue is dysplastic and results in tricuspid regurgitation

History

Wilhelm Ebstein (1864): Wrote the first description of 'Ebstein's anomaly' based on clinical and necropsy findings of a 19 year old patient.
Yater and Shapiro (1937): Described ECG and CXR findings of Ebstein's anomaly.
Uhl's anomaly: Different from Ebstein's anomaly. It is almost total absence of myocardium of RV

Incidence:

1: 20000 live birth
0.3 – 0.7 % of all CHDs
40 % of all congenital malformations of TV
One case report of Ebstein's anomaly involving both TV and MV in a patient.

Equal incidence in both genders
Maternal intake of Lithium carbonate increases incidence of Ebstein's by 28 fold in siblings

Ebstein's anomaly: Pathology

TV: Normal Anatomy (See "The Tricuspid Valve Complex, page xi)

Normal Tricuspid valve
- 3 cusps
  - Anterior (ATL): Largest; semicircular / quadrangular
  - Posterior (PTL): Small. It is the only leaflet that is scalloped.
  - Septal (STL): As name implies, is attached to IVS. A part is also attached to posterior wall of RV.

TV leaflets in Ebstein's anomaly: Dysplastic leaflets

Septal leaflet:
- Typically rudimentary. May be thickened and dysplastic.
- Displaced apically due to distal attachment of its basal part.
  - Displacement may be true (valve tissue cannot be identified on inlet septum) or false (leaflet attached to the inlet septum rather than displaced).
  - Impaired mobility: Either due to short chordae tendinae which tether the valve to IVS or due to valve itself (thick, nodular and fibrotic)

Posterior leaflet:
- Thickened and has abnormal chordal attachments.
- Apically displaced like STL. The maximum displacement is at the level of commissures between septal and posterior leaflet.
- Impaired mobility like STL.

Anterior leaflet:
- Almost always it is normally attached to the atrioventricular junction.
- "Dysplastic but not displaced".
- It is large, redundant and mobile. Rarely mobility is impaired due to short chordae tendinae that tether the leaflet to ventricular wall.
- It has fibrous and muscular strands running throughout the valve tissue as compared to only fibrous strands in normal heart.
- Attached to the ledge between inlet and trabecular zones of RV by chordae tendinae or attached to anterior papillary muscle.

Anatomy of right sided cardiac chambers in Ebstein's Anomaly:

In Ebstein's anomaly due to abnormal development of TV leaflet, the leaflets (especially the septal) are displaced apically into the RV. It leads to formation of 3 morphologic parts of right sided cardiac chambers.

RA proper: RA which would be there if the leaflets were properly attached.

Functional RV: Formed by the distal trabecular and outlet (infundibular) part of RV. There may be absolute decrease in number of muscle fibers throughout the RV. This may lead to dilatation of RVOT; seen in some cases of Ebstein's anomaly.

Atrialized RV: The intervening portion of RV (anatomically) between RA proper and functional RV. This part is thin walled, devoid of muscular tissue and is dilated. It contracts with RV. On electrophysiological studies the potentials obtained from this part are ventricular potentials. However, due to displaced leaflets it is anatomically continuous with RA.

The relative size of chambers depends upon the severity of displacement of valve. More apical the displacement; larger is the atrialized RV and smaller the functional RV.

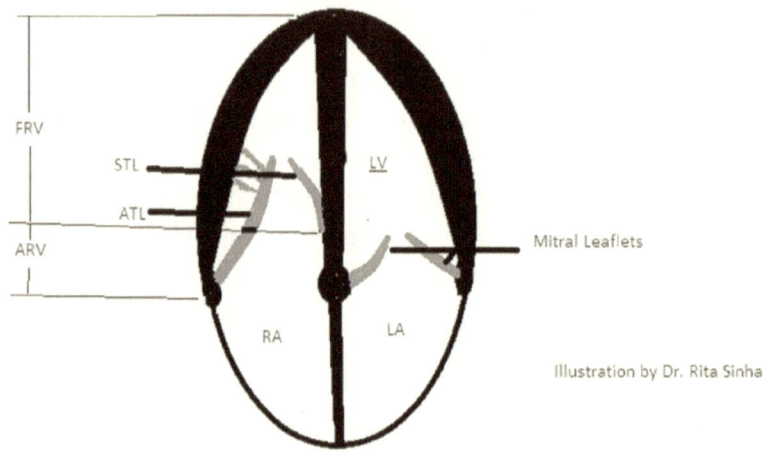

Illustration by Dr. Rita Sinha

LV in Ebstein's anomaly:
    In Ebstein's anomaly, primarily due to IVS motion and displacement there is reduction in resting LV function. LVEF increases in response to exercise.
    Left ventricular function is adversely affected by
        Diastolic position of IVS (leftward displacement) leading to reduction in LV diastolic volume
        Paradoxical motion of IVS
        Increase in fibrous tissue in LV free wall

Associated malformations in Ebstein's anomaly:

    PFO: In majority of cases
    Ostium Secundum ASD: In about 30 % cases.
    Mitral valve prolapse
    Other ASD: Primum or coronary sinus ASD.
    PAPVC
    Pulmonary valve stenosis or atresia

VSD / PDA
TV atresia
Coarctation of aorta
Right aortic arch
TOF

    Syndromes:
        Marfan syndrome
        Down's syndrome
        Bonnevie – Ulrich syndrome
        Cormelia de Large syndrome
Conduction system: Normal

Coronary system: Normal. High incidence of left or co-dominance of RCA and LCx.

Pathophysiology

TV morphology: Most commonly leads to incompetent valve of various degrees of severity. Rarely, it can be stenotic or atretic.

Ratio of atrialized to functional RV: Higher the ratio larger is the atrialized part of RV. Atrialized RV is nonfunctional, asynchronous with RA and interferes with RV filling. Even the functional RV does not show normal function due to less number of myocardial fibers.

Clinical features of Ebstein's anomaly

Symptoms:

    Cyanosis:

        Transient:
            Neonatal cyanosis: Present in more than 50 % cases, usually disappears and symptoms recur after first decade. In severe Ebstein's and with associated malformation, the prognosis is grave.
            High PVR (pulmonary vascular resistance) at birth leads to high RV systolic pressure (RVSP) leading to TR and high right atrial pressure (RAP). It causes right to left shunt across PFO or ASD, leading to cyanosis. Gradual decrease in PVR after birth causes consequent decrease in RVSP and RAP and the right to left shunt ceases. Later if there is development of RV failure, RAP increases again and with further increase in severity of TR, leads to right to left shunt across PFO or ASD. There is reappearance of cyanosis (bimodal presentation).
            Rare cases where the TV is stenotic or atretic,

        the RAP is higher than the usual cases. In such cases there is persistent cyanosis and raised JVP.
    Adult age: Transient cyanosis can also be seen during arrhythmias (e.g., PSVT). These patients are prone to arrhythmias

  Intermittent or persistent cyanosis:
    Occurs as a bimodal presentation in later part of life. Cyanosis is out of proportion to other symptoms and good effort tolerance despite cyanosis is diagnostic of Ebstein's anomaly. Cyanosis once established is usually progressive.

Dyspnea:
  Low cardiac output.
  Arrhythmias

Fatigue:
  Low cardiac output

Syncope:
  Common symptom. Occurs mainly during arrhythmias.

Palpitations:
  Common.
    Atrial arrhythmias due to atrial fibrillation or flutter are very common.
    Accessory pathways (WPW syndrome) are common in Ebstein's.
      Usually it is right sided free wall pathways and could be multiple.

Chest pain:
  Of obscure origin.
  Sharp/stabbing or shooting in quality
  Can occur in right or left anterior chest / epigastric or

retrosternal areas

According to some it is due to hemorrhagic pericarditis, probably related to blood leaks from atrialized RV into the pericardium.

Right heart failure:
Usually it is the terminal event.

IE: Rare

Paradoxical emboli and brain abscess

Signs:

Growth/development/body built:	Normal
Cyanosis:	Present in 50 – 80 % cases
Clubbing:	Varying degree of clubbing is associated with cyanosis
Pulse:	Usually normal. Rarely low volume due to reduced LV stroke volume
JVP:	Normal;

* if JVP is raised: Reconsider the diagnosis of Ebstein's anomaly.

In spite of severe TR the A and V waves and X descent are attenuated due to the damping effect of the commodious RA and thin, toneless atrialized RV. The low pressure TR ensures normal JVP. In rare case of RV failure the mean JVP is raised.

Cardiovascular Examination:
Inspection: Precordium: Normal
Rarely left parasternal prominence and very occasionally right anterior chest prominence due to large RA.
Systolic undulating or rippling motion is **seen (but is not palpable)** along mid to lower left sternal border (LSB), especially with breath held during inspiration: Due to ARV

Palpation:
- Apex: LV type
- Pulsation overlying pulmonary trunk is always absent:
  - *If present suspect diagnosis other than Ebstein's.*
- Systolic impulse in 3rd left intercostals space: Due to infundibular part of RV
- Systolic thrill: Absent except in some neonates prior to fall in PVR.
  - *If present consider diagnosis other than Ebstein's anomaly.*
- Epigastric pulsations: Absent in almost all cases except in some symptomatic neonates;
  - *If present consider diagnosis other than Ebstein's anomaly (e.g., organic TV disease)*

Percussion:
- Enlarged right heart border due to RAE

Auscultation:
- **Sounds**: Characteristic feature: "Multiplicity of sounds"
  - **S1**: Distinct M1 and T1 component
    - *T1 best heard between apex and lower left sternal edge
  - Widely split S1:
    - Due to delayed closure of TV:
      - Due to large size and increased incursion of ATL it requires longer time to reach a tense and fully closed position
      - RBBB
      - Hypokinetic RV
  - Loud T1:
    - Loud T1 ("Sail sound"):
      - Due to increased tension developed due to increased excursion of ATL
    - "Sail sound": Character is like a 'click sound' produced due to 'closure' of TV

Single S1: Only T1 is heard
- Soft M1
  - Long PR interval softens M1
- Muffled M1
  - Due to pre-excitation, an early T1 is produced which buries M1

S2 Variable
- Widely split: Delayed P2 due to RBBB
- Wide and fixed split: Associated ASD; however unlike other causes of fixed split in these cases P2 is soft due to decreased pulmonary flow.
- Paradoxical split: Pre-excitation syndrome
- Single S2: Inaudible P2 due to low pressure and low pulmonary flow

S3 / S4
- Common
- Produce distinct triple or quadruple rhythms
- Louder in inspiration because of RV origin
- Both sounds summate if associated with prolonged PR

Opening snap
- TV origin
- Opening movement of ATL

Murmurs
- Systolic murmur of TR:
  - Murmur due to TR may range from absent to loud
  - Best heard at a site lateral to conventional tricuspid area (towards apex) because of displaced TV
  - Does not increase with inspiration (Compare with organic TV disease) because Functional RV cannot increase the stroke volume and regurgitant volume any further.

Duration: Systolic to holosystolic
Length: Decrescendo and ends before S2
Character: Scratchy and harsh. Medium frequency murmur due to normotensive TR.
Associated with diastolic murmur.

MDM
Diastolic murmurs in Ebstein's anomaly: heard due to
Summation of RVS3 and RVS4
Associated TS
Flow murmur due to severe TR

Investigations

1. ECG

Always abnormal

P wave
- Characteristically abnormal in height, duration and shape due to abnormal and prolonged conduction in large RA
- If p wave's height is more than R wave height in the same lead; called as HIMALAYAN P wave (seen in one-third cases; described by Helen Taussig)
- Progressive increase in P wave height correlates with clinical deterioration and poor prognosis
- Absent 'p' wave: permanent atrial standstill: Rare but seen in familial Ebstein's anomaly

PR interval
- May be normal; prolonged (due to increased size of RA) or short (due to associated WPW syndrome)

QRS:
- Wide
    - Complete RBBB (75 – 95 %)
        - Due to abnormal and prolonged conduction in ARV.
        - R' wave is never more than 7 mm because there

is no RVH in Ebstein's anomaly

"Splintered QRS": Seen in 1/3 rd cases.
- Distinctive pattern of bizarre '2nd' QRS complex attached to preceding normal QRS;
- Produced due to ARV

Axis: Normal; abnormal if associated with WPW syndrome (LAD)
- Deep Q wave in leads II, III and aVF (Associated WPW syndrome; posteroseptal accessory pathway)
- Q wave in V1 (50 %) and extending up to V4 (D/D ASMI)
- Caused by RV intra-cavitary potentials recorded due to leftward displacement of TV)

Standard limb voltage < 7 mm

WPW syndrome: Seen in 5 – 25 % cases of Ebstein's anomaly (Only CHD which is consistently associated with preexcitation)

Rhythm
- Sinus rhythm
- Arrhythmias
    - Narrow QRS tachycardia (25 – 30%)
        - PSVT
        - Atrial arrhythmias: Atrial fibrillation, Atrial flutter or tachycardia
    - Wide QRS tachycardia
        - Antidromic tachycardia due to WPW syndrome (5-25 %)
        - Right free wall
        - Posteroseptal pathways
        - Multiple pathways in 1/3 rd cases
        - Rarely due to Mahaim fibres

2. CXR

Pulmonary blood flow: Normal to reduced

RAE: Most consistent feature of Ebstein's. Present even if CE is absent.

Narrow vascular pedicle: Decreased forward cardiac output; Ebstein's anomaly is the only cyanotic CHD where both aortic root and pulmonary trunk are likely to be normal or reduced in caliber

D/D
- *TGA: due to anteroposterior relation of great vessels*
- *SV*
- *Pericardial effusion*

Cardiomegaly: Consistent feature in symptomatic patients

Straight left upper border
- Due to aneurysmal dilatation of RVOT

BOX / FUNNEL shaped heart: Due to RAE and aneurysmal RVOT

D/D:
- *TAPVC*
- *Tricuspid Atresia*
- *SV*

Summary:
- CXR in Ebstein's
  - Cardiomegaly
  - Narrow pedicle
  - Normal to decreased pulmonary flow
  - Small PA (D/D MR with PH; ASD with PH)
  - Inconspicuous SVC shadow
    - Pleural effusion with prominent SVC shadow: Think of RVEMF not Ebstein's anomaly

3. Echocardiography

2 D Echo
Aneurysmal RA
TV:
Incomplete coaptation of leaflets
STL:
Apical displacement
8 mm / m² in systole
Absolute value
> 20 mm: Adults
> 15 mm: Children
Mitral – apex distance / Tricuspid – apex distance
Normal: 1.0 – 1.2: 1.0
Ebstein's: > 1.8 – 3.2; 1.0
ATL:
Attached to annulus
Long and thickened
Redundant
May be tethered to RV wall
Restricted mobility (whip or sail like motion)
Diastolic doming may be present

Large ARV
Small FRV
Normal LVEF; Decreased RVEF
Aneurysmal RVOT

M Mode
Simultaneous recording of TV and MV with a transducer position outside left mid clavicular line is a clue to **diagnosis** of EA
Abnormal late closure of TV (> 50 msec) is **diagnostic** of Ebstein's anomaly
Normal McTc: 0 - 0.03s (Farooki et al) / -0.005 to 0.05 s (Milner et al)
So according to Farooki et al: McTc delay > 0.03s

and according to Milner et al McTc delay > 0.065s is specific for EA

Daniel et al (Br. Heart J 1980; January; 43(1): 38–44) supports > 0.065s criteria for diagnosis of Ebstein's anomaly.

ATL

Anterior tricuspid leaflet opening amplitude (OA) twice that of anterior mitral leaflet is s/o EA.

EF slope flat or decreased

Larger rectangular TV echo and dwarf smaller MV echo

RVVO

Paradoxical IVS motion

Small LV

Premature opening of PV due to large forceful RA contraction

Color – Doppler study

Severe normotensive TR: Due to normotensive flow, the color of TR jet may not be impressive

Limitation of echocardiography in Ebstein's anomaly:

Following features are not well recognized

Morphologic features of posterior TV leaflet

Apical RV long axis views and PSAX are better views to visualize PTL.

Cleft, perforation or fenestration of ATL

Shina and Tajik Scoring system: (Circ 68, No. 3, 1983; 534-44)

A very useful scoring system based on 2 D features to plan type of surgery once intervention is advised in Ebstein's anomaly. The decision of surgical intervention is not based on echocardiographic findings but depends on symptom class of patients. Once a surgical intervention decision is taken, the following features help in deciding about type of intervention.

Feature	Score
Severe ATL tethering	3
Restricted ATL motion	2
Mild ATL tethering	1
Functional RV < 35 % (FRV / TRV)	2
Absent STL	1
Displaced ATL	1
Aneuysmal RVOT	1
RA > 60 mm / m2 in diameter	1
Severe TVP	1
Total score : 13	

A score more than 2 was highly indicative of a patient needing surgery. An index number of 5 or more in this study was highly indicative of the need for valve excision. Value less than 2 was found in both groups of patients who were advised either medical therapy or valve repair.

Of particular importance was the immobilization of the anterior leaflet which reduced the likelihood that the surgeon would be able to construct a competent unicuspid valve by plastic repair. Tethering and immobility of ATL and a small FRV were the best indicators of patients who required valve excision in study by Shina et al. *As stated earlier, this echo score does not classify the group of patient who will require surgery. The decision of surgery is primarily based on the functional class of patient.*

In conclusion, 2 D echocardiography allows excellent preoperative assessment of morphological features of Ebstein's anomaly. These observations can be used to measure morphologic severity and, when indexed, can accurately determine which patient is likely to require valve excision.

Measurements
TV annulus: Maximum diastolic dimension across RV inlet at the level of TV annulus
Long axis RV: TV annulus to RV apex in diastole

ARV: Maximal distance from TV annulus to leading edge of TV leaflets in systole
FRV: RV − ARV
Index of anatomic severity: FRV / RV X 100
RA: Maximum systolic distance from TV annulus to posterosuperior wall of RA
STL displacement: Measured in systole as distance from tricuspid annulus to the nearest point of septal attachment.
Aneurysmal RVOT: RVOT anteroposterior dimension greater than 2 times aortic dimension)

4. Cardiac Catheterization

Preferably should be avoided due to life threatening arrhythmias
Electromechanical events recorded by Zucker catheter
RA angiogram:
  "Trilobed appearance" due to
    Aneurysmal RA
    Smooth walled ARV
    Trabeculated FRV
      These are separated by 2 notches in AP view:
        Proximal notch is due to TV annulus and distal notch is due to displace TV
      'Teeter − Totter' appearance: Alternate expansion and contraction of dilated right atrial appendage.

RA angiogram: Diagnostic findings are seen in
  Ebstein's anomaly
  Tricuspid atresia
  Constrictive pericarditis
  RVEMF

C TGA with Ebstein's anomaly of Left AV valve: Characteristics
  Not loud as ATL is small
  TV: stenotic / regurgitant
  ATL: small, cleft (30%)
  ARV: Neither so thin nor striking
  Left sided WPW syndrome
  Annulus not dilated
  Always requires valve replacement

Natural History

Hamish Watson

An international co-operative study of 505 cases
- N: 505
- Men: 258
- Women: 247
- Age distribution
  - Less than 1 yr of age: 35,
    - 72 % patients were in CCF
  - 1 –25 yrs: 403
    - Largest group
    - 71 % of cases were either in NYHA class I or II.
  - More than 25 yrs.: 67 patients;
    - 60 % of cases were in NYHA class I or II.

Cardiac Catheterization was performed in 363 patients.
13 patients died during the procedure and 6 had cardiac arrest.
- 90 patients had SVT during the procedure.
- 48 % cases had associated malformation; however if interatrial septal defects were excluded; only 12 % had associated malformation.

Necropsy was done in 93 cases.
48 % had associated malformation and if IAS defects were excluded 8 % had associated malformations.

Death: 67 (13.3 %) died during follow up.
Surgical intervention was the most common cause of death (> 50 % cases). The surgical experience suggest that palliative procedures have little or nothing to offer and that the replacement of malformed TV should be delayed if possible, until the patients' heart is large enough to take an adult size prosthesis or homograft.

Natural History: AHJ 1993

Salient points:
This study showed 2 distinct groups of patients. Patients with or without associated malformation in Ebstein's anomaly of TV.

50 % of patients with Ebstein's anomaly die during infancy. Those who survive infancy, another 40 % die in their first decade of life. Only 60 % reach adulthood and most of them are in NYHA class I or II.

Sudden cardiac death is seen in 2.6 % cases.
The patient group without associated malformation shows a better prognosis. 70 % of such patients survive infancy.
Operative mortality is high in younger patients.

Sudden cardiac death: SCD can occur in Ebstein's anomaly regardless of severity of disease. More common in 5th decade.

Poor prognostic factors
- Diagnosis during infancy:
    - Those who survive infancy have a relatively better prognosis if cyanosis or arrhythmias are absent.
- NYHA class III or IV
- Cardiomegaly > 60 % of cardiothoracic ratio (CTR)
- Presence of cyanosis: After onset of cyanosis disability tends to be progressive
- Recurrent PSVT / Atrial flutter or fibrillation
- The advent of chronic atrial fibrillation predicts death within 5 yrs.
- RV failure
- Progressive increase in P wave height and decrease in QRS voltage
- Associated malformation other than IAS defects
- Increased hematocrit
- Increased right atrial pressure (RAP)
- Cardiac surgery

Management

Recommendations for surgery in neonates and pediatric patients for Ebstein's anomaly with severe tricuspid regurgitation

Indication Class

1. Unstable cyanotic newborn in congestive heart failure, in need of mechanical ventilation, prostaglandin dependent and failed medical therapy I B

2. Congestive heart failure I B

3. Deteriorating exercise capacity (New York Heart Association functional class III or IV) I B

4. Progressive cyanosis with arterial saturation <80% at rest or with exercise I B

5. Asymptomatic patient with increasing tricuspid insufficiency and cardiothoracic ratio II C

Recommendations for surgery in adolescents or young adults with Ebstein's anomaly and severe tricuspid regurgitation

Indication Class

1. Congestive heart failure I B

2. Deteriorating exercise capacity (NYHA functional class III or IV) I B

3. Progressive cyanosis with arterial saturation <80% at rest or with exercise I B

4. Progressive cardiac enlargement with cardiothoracic ratio >60% IIa C

5. Systemic emboli despite adequate anticoagulation IIa C

6. NYHA functional class II symptoms with valve probably reparable IIa C

7. Atrial fibrillation IIa C

8. Deteriorating exercise tolerance (NYHA functional class II) IIa C

9. Asymptomatic patients with increasing heart size IIb C

Contraindication

Asymptomatic patients with stable heart size III C
NYHA New York Heart Association
* TR severity and ASD size and pattern of shunt are not criteria for surgery

Summary:

Class I recommendations for surgery in neonates and pediatric patients with Ebstein's anomaly and severe tricuspid regurgitation

- Unstable cyanotic newborn with congestive heart failure, in need of mechanical ventilation, prostaglandin dependent and who has failed medical therapy;

- Congestive heart failure;

- Deteriorating exercise capacity (NYHA functional class III or IV); or

- Progressive cyanosis with arterial saturation less than 80% at rest or with exercise.

Class I recommendations for surgery in adolescents or young adults with Ebstein's anomaly and severe tricuspid regurgitation

- Congestive heart failure;

- Deteriorating exercise capacity (NYHA functional class III or IV); or

- Progressive cyanosis with arterial saturation less than 80% at rest or with exercise.

Surgical Options:

    Palliative
        Glenn: SVC to PA shunt to increase pulmonary blood flow
            Fontan
    Definitive: Consists of
        Right atrial reduction
        Atrialized ventricle (ARV) plication
        TV: Either repair or replacement
        Correction of other association malformation
            ASD closure (may be avoided to prevent low flow state in post operative period)
            VSD closure
            TS: Valvotomy
            WPW syndrome: Radiofrequency ablation or surgical division

Other options
- Neonatal Ebstein's: Convert to tricuspid atresia with AP shunt and Fontan at a later date
- Cardiac transplant

Type of surgery: *Surgical treatment for Ebstein's anomaly has 2 major options, tricuspid valve replacement and various types of plastic reconstruction or repairs*

Repair: Successful plastic repair depends on a mobile, normal sized or enlarged ATL.

- Danilson repair
  - Pre requisites:
    - Elongated ATL
    - No restriction of ATL movement: It can function as monocusp valve
  - Technique:
    - Plication and marsuplization of ARV
    - TV annuloplasty if TV annulus is > 4.0 cm
    - Early and late mortality: 7 %

- Carpenter repair:
  - Aim: Reconstruct TV and RV
    - Plication of ARV:
    - Decreases chances of thrombosis
    - Increases atrial contractility

- Hardy repair:
  - Plicate and excise RA along with usual technique

Tricuspid valve replacement:
- Pronounced tethering of ATL and small FRV (< 35 %) were the two most important factors why patients required replacement rather than repair.

Mortality:
  Overall: 20 %
  Early: 5 %
  Late: 15 %
TV replacement is not curative and does not decrease risk of SCD

Scoring system for risk of surgery and mortality (JACC 1992)

Depends on the ratio of RA + ARV / RV + LA + LV

Ratio	Grade	Mortality
< 0.5	I	0 %
0.5 – 0.99	II	10 %
1.0 – 1.49	III	44 %
> 1.5	IV	100 %

Pulmonary Valve

It lies slightly anterosuperior to the aortic valve at the superior end of the infundibulum of the right ventricle; this is roughly the level of the left third costal cartilage. Its plane faces posterosuperior and to the left. The diameter of the valve is 2-3cm.
Like the aortic valve, the pulmonic valve is formed by 3 semi lunar leaflets or cusps.

Each valve cusp is characterized by:
- A central layer of collagen, the lamina fibrosa
    - A thickening of the lamina fibrosa at the free margin of each leaflet - the nodulus. Regions lateral to the nodulus of each valve are termed the lunules
- A layer of endocardium completely covering the valve
- An attachment between the lamina fibrosa and the pulmonary valve annulus, part of the fibrous skeleton of the heart
- A dilation of the wall of the pulmonary trunk immediately superior to each cusp - the sinuses of Valsalva; each sinus fills with blood, so preventing the valve leaflet from adhering to the wall of the pulmonary trunk with consequent valvular incompetence

In contrast with the aortic valve, the cusps of the pulmonic valve are supported by freestanding musculature with no direct relationship with the muscular septum; its cusps are much thinner and lack a fibrous continuity with the anterior leaflet of the right atrioventricular (AV) valve (tricuspid valve)

The cusps of the pulmonic valve are defined by their relationship to the aortic valve and are thus termed anterior or nonseptal, right and left cusps. They can also be defined by their relationship to a commissure found in the pulmonic and aortic valves and hence termed right adjacent (right facing), left adjacent (left facing), and opposite (nonfacing).

Pulmonary Stenosis (PS)

Etiology

Congenital: Most common cause
RHD: Very rare;
    Never an isolated involvement: Usually with quadrivalvular involvement
Malignant carcinoid

Congenital PS

Incidence:
    Can occur either in isolation or in association with other defects in 20-30 % of patients with CHD
    Commonest obstructive congenital heart disease

Pathology:
    Can have one or more of the following features
        Fusion of valve leaflets with mild degree of valve thickening
        Abnormality of cusps: Bicuspid, quadri or unicuspid valves
        Dysplastic valves: Markedly thickened and immobile cusps with minimal fusion
            Seen in only 10 -15 % cases of PS
            Familial pattern
            Seen in Noonan's Syndrome

D/D between 'dome shaped' PS from pulmonary valve dysplasia

Feature	Dome shaped valve	Dysplasia
Commissural fusion	+	-
Bi or quadricuspid valve	+/-	-
Thickened & immobile cusp	-	+
Positive family history	-	+
Associated maldevelopment	-	+
EC and audible P2	+	-
Pulmonary regurgitation	-	+/-
Post stenotic dilation of PA	+	-
Hypoplastic PV annulus	-	+

+: Present, -: Absent, +/-: may or may not be present

Cardiac lesions associated with valvular PS
- ASD and PFO
- VSD
- TOF
- Hypoplastic RV
- Pulmonary artery stenosis (in maternal rubella syndrome)
- Supravalvular aortic stenosis (William's syndrome)

Clinical Features

Symptoms:

    Abnormal facial appearance
        Noonan's syndrome
        William's syndrome
        Alegille's syndrome (Arteriohepatic dysplasia)
        Maternal rubella syndrome
    Growth and development: **Normal as a rule**
    Dyspnea and fatigue (result from decreased cardiac output)
    Precordial pain
    Dizziness, Syncope and sudden death
    Peripheral edema
    Cyanosis:
        Almost always due to right to left shunt across IAS
        Not a reliable criterion for severe PS
    Squatting: Extremely rare; should **revise** the diagnosis of isolated PS if squatting is present
    Hemoptysis: Hemoptysis in a suspected case of PS should always point to **peripheral pulmonary stenosis.**

Factors leading to progression of symptoms in PS

    Development of infundibular PS due to generalized RV hypertrophy

Reduction in valve area index due to failure of orifice growth to keep pace with body growth in children
Excessive chronic RV pressure overload with decompensation
Chronic RV myocardial ischemia even in absence of CAD
Pulmonary valve calcification (rare)

Signs:

JVP

Feature	Findings	Mechanism / significance
a. Level	Normal or elevated	Elevated with RV failure
		Associated organic TS
b. Waves		
A wave	Prominent	Prominent A wave with severe PS; It also indicates intact IVS in a case of PS
		Associated organic TS
		May merge with 'v' wave
X descent	Normal	Obliterated if associated with TR
V wave	Normal	Increased with RV failure
Y descent	Normal	Rapid with RV failure or TR

Precordial palpation
> RV systolic impulse:
> Should always be present; if **absent revise** diagnosis of isolated PS
>> A prominent pulsation in the second left intercostal space is **never** felt in valvar PS in spite of post stenotic dilation. Because elevation of the pressure in MPA is necessary to produce such pulsation, its presence indicates peripheral pulmonary rather than valvar stenosis.

Auscultation:

> S1: Normal

> S2:
>> P2 : Soft
>> Wide split
>> Wide and fixed: Severe PS due to fixed RV stroke volume

> S3: Absent; if present suspect associated ASD or PAPVC

> S4: Frequently present: if **absent, revise** diagnosis of significant PS

> EC: Present; it is the **only** right sided event whose intensity decreases with inspiration and increases with expiration

> EC is **absent** in
>> Severe PS,
>> Dysplastic PS,
>> Infundibular PS and
>> PPS

Murmur:
    ESM: Best heard at upper left sternal border
        Well conducted to entire precordium, **neck** and back
        Systolic thrill: **Pathognomonic** sign of PS; the intensity of thrill is **directly** proportional to severity of stenosis.

Clinical criteria of severe PS

1. P2:
   Intensity of P2 is **inversely** proportional to severity
   Wide split S2 (> **100 msec**) indicates severe PS
   Wide and fixed split S2 indicates severe PS (fixed RV stroke volume)
2. EC:
   Absent in severe PS
   Interval between S1 and EC is **inversely** proportional to severity
3. Length of ESM is **directly** proportional to severity of PS. Loudness is not a criterion however it is unusual to find less than grade 3 murmur with severe PS.
4. Shape: Severe PS is late peaking murmur or 'kite' shaped murmur
5. RV S4 with large 'a' waves or 'presystolic' hepatic pulsations
6. In infants: These classical signs are not present. Presence of cardiomegaly, pansystolic murmur of TR and very 'soft' P2 are suggestive of critical PS
7. Symptom: Cyanosis

Investigations

1. ECG:

With **few** exceptions, a reasonable estimate of severity of PS can be made by an appropriately obtained ECG

ECG findings of severe PS:
ECG is almost always abnormal in severe PS
QRS axis: + 110 – 160° or more
    Dominant 'R' in aVR or V1
    V1 may show Rs, pure R or qR (tall R) pattern
        qR in V1 indicates RAP > 8 mm Hg
    Height of R in V1 and V4R correlates with RV systolic pressure at rest (Between 2 – 20 yrs of age)
    Ht of R (in mm) X 5 = RV systolic pressure in mm Hg
No abrupt transition of QRS in the pattern of QRS in mid precordium
Discordant 'T' wave from V4R to V4: If present indicates RVSP > 100 mm Hg
ST depression beyond V2
RAE: Abnormally tall and peaked 'P' waves in lead II or right precordial leads. It is an insensitive sign of severe PS.

Left axis deviation (LAD) in a case of PS: Causes
- Hypoplastic RV (relative LAD of + 30 to + 70 degree even in presence of severe PS)
- Noonan's syndrome (LBB conduction defect): Also associated with counterclockwise loop (CCL)
- Rubella syndrome: Indicates associated PPS: CCL may also be present.
    - It is due to myocardial damage from the rubella virus
- Coexisting LBB conduction defect may be present as an isolated defect
- Supravalvar PS: LAD with CCL

2. CXR

Prominent MPA and LPA:
- **Almost invariably** found with significant PS.
    - (Abram and Wood reported 100 % incidence but it has not been confirmed by others.)
- The degree of dilation does **not** correlate with severity.
- MPA dilation is **absent** in
    - Dysplastic PS
    - Severe infundibular PS
    - Bilateral peripheral PS with valvar PS
    - Rubella syndrome: Valvar with supravalvar PS
- **Normal** pulmonary vasculature
    - Oligemia is seen with associated RHF or ASD
- Dilated right sided cardiac chambers: seen with severe PS
    - RAE: In 50 % cases
    - RV enlargement seen with RHF (end stage disease)

3. Echocardiography:

M mode
- An increased "a" dip ($\geq$ 8 mm) of the pulmonic valve
- Right ventricular hypertrophy (wall thickness > 9 mm)
- Increased right ventricular ejection time

Prominent a dip on m mode PV

Prominant 'a' wave dip on M Mode PV

2 D echo:
    Decides about valvar, subvalvar or supravalvar site of stenosis with help of Doppler studies
    Doming of PV with restricted systolic motion
    Dilation MPA and branches especially LPA; In ASD the dilation usually involves RPA.
    Differentiates with dysplastic PS
    PV annulus size should be measured to plan for balloon size for BPV
    Atrial level shunts and coexisting anomalies can be diagnosed

Color Doppler studies
    Severe PS:
        Peak gradient > 70 mm Hg across PV is diagnostic of severe PS
        RV systolic pressure ≥ LV systolic pressure is severe PS

    Mild PS:
        Peak gradient < than 40 mm Hg
        RV systolic pressure less than half of LV pressure

    Moderate PS:
        Peak gradient 40 - 70 mm Hg
        RV systolic pressure 50 – 75 % of LV systolic pressure

4. Cardiac catheterization

In present era, catheterization is mainly required for intervention (BPV)
It clearly demonstrates severity and location of stenosis; best seen in lateral view.
RVEDP is elevated only with RV failure

Assessment of severity of PS (Keith et al)

1. Catheterization

Severity	Gradient (mm Hg)	RVSP (mm Hg)
Mild	< 50	< 70
Severe	> 80	> 100

2. At least one of the following

   Cyanosis
   CHF
   ECG: 'S' in lead I $\geq$ 15 , 'q' in V1 or T inversion in aVF with R V1 + S V6 $\geq$ 35 corresponds to gradient of 110 mm Hg across PV

3. 3. $\Delta P = RVSP - PASP = (10.5 \times ISM) + (2.6 \times S \text{ in lead 1}) + S2 \text{ score} + T \text{ score}$

   Where ISM: Grade of ESM (from grade 1 – 6)
   S in lead 1: in mm (1 mm = 0.1 mV)
   S2 score:
       If P2 is normal: -10,
       If P2 is audible but soft: +2 and
       If P2 is absent: +15
   T score: + 15 if T is biphasic in V1 when R in V1 > 10 mm

If estimated Δ P is < 35 then the PV gradient on cath was less than 65 mm Hg and if Δ P was > 50 then it corresponded to severe PS.

Management

Options:

> Balloon Pulmonary valvotomy
> Surgery
> > Surgical valvotomy
> > Valve replacement

BPV is the treatment of choice

Indications for Balloon Valvotomy in Pulmonic Stenosis

Recommendations for intervention in children with pulmonary stenosis (balloon valvotomy or surgery)

Indication Class

1. Symptomatic infant with critical pulmonary stenosis I B

2. Patient with NYHA class III to IV (exertional dyspnea, angina, syncope or presyncope) and critical pulmonary stenosis I B

3. Asymptomatic patient with normal cardiac output (estimated clinically or by catheterization)

a) RV-PA gradient >50 mmHg I B
b) RV-PA gradient 40 to 49 mmHg IIA C
c) RV-PA gradient 30 to 40 mmHg IIB C

Contraindication

4. Asymptomatic patient with normal cardiac output (estimated clinically or by catheterization) with RV-PA gradient <30 mmHg III C

NYHA New York Heart Association; RV-PA Right ventricular to pulmonary artery

Recommendations for intervention in adolescents or young adults with pulmonary stenosis (balloon valvotomy or surgery)

Indication Class

1. Patients with exertional dyspnea, syncope, or presyncope I B

2. Asymptomatic patients with normal cardiac output (estimated clinically or determined by catheterization)

    a) RV-PA peak gradient >50 mmHg I B
    b) RV-PA peak gradient 40 to 49 mmHg IIa C
    c) RV-PA peak gradient 30 to 39 mmHg IIb C

Contraindication

2. d) RV-PA peak gradient <30 mmHg III C

RV-PA Right ventricular to pulmonary artery

Summary:

Class I recommendations for intervention in children with pulmonary stenosis are as follows

- Symptomatic infants with critical pulmonary stenosis;

- Patients with NYHA III to IV (exertional dyspnea, angina, syncope or presyncope) and critical pulmonary stenosis; or

- Asymptomatic patients with normal cardiac output, estimated by echocardiography or by catheterization (RV to pulmonary artery [RV-PA] gradient greater than 50 mmHg).

Class I recommendations for intervention in adolescents or young adults with pulmonary stenosis

- Patients with exertional dyspnea, syncope or presyncope; or

- Asymptomatic patients with normal cardiac output, by

  Echocardiography or determined by catheterization (RV-PA peak gradient greater than 50 mmHg).

Balloon Pulmonary Valvotomy

BPV is effective in cases where
PV Annulus is more than 75 % of the predicted annulus for the age and BSA
Valvar PS (not associated with infundibular PS)
Non dysplastic valve

Balloon size should be at least 10 – 20 % higher than the annulus size
If annulus is more than 20 mm then double balloon technique is more effective

Inoue balloon has been used effectively to treat PS

Surgical valvotomy:

    Only if BPV is not possible or unsuccessful
    If preoperative gradient is more than 200 mm Hg, infundibulectomy is almost invariably necessary

    Valve replacement is very rarely required; required usually when annulus is also hypoplastic.

PS vs PPS (Peripheral PS)

The following points favor PPS:

    Hemoptysis in a suspected case of PS is usually due to PPS
    P2: Either normal or mildly increased; rarely very loud
        P2 in PPS is typically normal not loud; because PH if present is usually systolic
    No EC
    ESM:
        Widely heard but not in neck
        Low intensity
    Rarely continuous murmur: (Murmur in PPS is typically systolic rarely continuous; because a diastolic gradient usually does not exist across the stenosis)

PA branch stenosis is defined as when a localized narrowing area is found within PA branches and smallest diameter of the narrow part is less than or equal to 50 % of the largest diameter of the ipsilateral artery. Differences in diameter between the distal part of RPA and LPA of less than 30 % are defined as imbalance of PA growth.

Pulmonary Regurgitation (PR)

Etiology

Secondary to pulmonary hypertension of any cause (Dilation of valve ring)
Infective Endocarditis
Idiopathic pulmonary artery dilation
Marfan's Syndrome or Connective Tissue Disorder
Congenital
Iatrogenic
Others:
    Trauma
    Carcinoid Syndrome
    RHD
    Syphilis

Clinical Features

Signs:

    Hyperdynamic RV
    Palpable systolic pulsation in left parasternal area and in pulmonary area.

Auscultation:

    S1: Normal
    S2: Normal to Wide split S2 (except when PH is present)
        P2: Variable
            If valve tissue is absent or reduced
                P2 is soft or inaudible
            If associated with PH: Loud P2
            Normal in most cases

    EC: Non valvular systolic ejection click
        Indicates significant dilation of MPA

    Murmur:

        Early Diastolic murmur
            Depends on size of the valvular defect and pulmonary arterial hypertension

Graham Steell's Murmur:
    EDM of PR due to severe pulmonary hypertension (PASP is at least 60 mm Hg)
    High pitch (resembles AR) and is decrescendo in shape beginning just after P2
If PR is not due to PH
    An early to mid diastolic crescendo-decrescendo murmur; medium pitch and ends well before S1

D/D with AR:
    Absence of peripheral signs of AR
    Presence of signs of severe PH

Investigations

1. ECG:

    Signs of pulmonary hypertension
    If PR is present without significant PH (causes other than severe PH)
        Signs of RVVO (rsr' or rsR' on ECG) are present

2. CXR:

    Prominent MPA
        Fluoroscopy: 'Hilar dance' due to increased RV stroke volume and diastolic collapse of pulmonary arterial tree
    Signs of PH
    RV is rarely prominent on CXR unless PV is absent or severe PH is present with PR

3. Echocardiography:

    Views used
        PSAX and modified PLAX
        Pediatric patients
            Subcostal long and short axis views

Usually abnormal with moderate to severe PR
- Length of PR jet
  - Less than 2 cm: Mild PR
  - 2 – 4 cms and occupies up to 50 % of RVOT: Moderate PR
  - \> 4 cms and occupies more than 50 % of RVOT: Severe PR
- RV volume overload
- Paradoxical motion of IVS (like any other cause of RVVO)
- Signs of PH

4. Cardiac Catheterization:

Rarely required
- Only if distinction from AR must be made or when associated lesions are suspected
- Characteristic pressure tracing of PR: **Near equilibration** of PA and RV end diastolic pressures is characteristics of significant PR

Management

Benign prognosis hence treatment is supportive in most cases.
Medical:
    Treatment of associated lesion
    Cardiac glycosides
Surgical: Rarely required
    Bioprosthetic valve is used in the rare event of replacement

Recommendations for pulmonary valve replacement in chronic severe pulmonary regurgitation

Indication Class

1. Ventricular tachycardia with moderate to severe pulmonary regurgitation I C

2. New onset tricuspid regurgitation with moderate to severe pulmonary regurgitation IIa C

3. Worsening New York Heart Association class with right ventricular dilation IIa C

Summary

Class I recommendations for pulmonary valve replacement in chronic severe pulmonary regurgitation

- Ventricular tachycardia with moderate to severe pulmonary regurgitation

Straight Back Syndrome

The straight back syndrome, consists of loss of normal upper thoracic spinal curvature. It is associated with cardiac murmurs. This entity, which was previously considered a form of 'pseudo heart disease', has been attributed to squashing of the heart in the reduced anteroposterior diameter of the chest

First described by Rawlings in 1960

> Rawlings also postulated that it was due to a congenital development defect affecting spine during the embryonic state (seen as early as 8th week of intrauterine life)

No radiological evidence of bone disease and vertebra are normal

Clinical features:

> Usually asymptomatic:
> > Diagnosed during investigations for ejection systolic murmur at pulmonary area
>
> Symptoms:
> > Palpitations
> > Chest pain
>
> Always asymptomatic for skeletal defect

Signs:
- Spinal deformity
  - Scoliosis and pectus excavatum are frequently seen
- Left parasternal pulsations
- ESM at pulmonary area
- Delayed P2
- Rarely MVP may be present

CXR findings

- Normal cardiac silhouette
- 'Pancake' appearance suggestive of cardiomegaly
- Leftward displacement of heart and prominent pulmonary arteries
- Loss of normal kyphotic curvature of thoracic spine

Measurements:

On lateral CXR

Distance between middle of anterior border of 8th vertebra to a vertical line drawn between 4th and 12th vertebra (normal value > 1.2 cm)
In straight back syndrome the distance is less than 1.2 cm

Distance between 8th thoracic vertebra to the posterior cortex of sternum (normal > 10.5 cm)
In straight back syndrome the distance is less than 10.5 cm

Distance between 4th intervertebral space (between 4th and 5th vertebra) to the sternal angle (Normal more than 8.6 cm)
In straight back syndrome the distance is less than 8.6 cm

Distance between 12th vertebra to the anterior costophrenic angle (normal more than 10 cm)
In straight back syndrome the distance is less than 10 cm.

Ratio of transverse and anteroposterior diameter:
≥ 3 is suggestive of straight back syndrome

Suggested Reading

1. Kitchin A, Turner R. Diagnosis and treatment of tricuspid stenosis. Br Heart J. 1964;26:354-379

2. Perloff PK, Harvey WP. Clinical recognition of tricuspid stenosis. Circulation. 1960;22;346-364

3. Gibson R, Wood P: The diagnosis of tricuspid stenosis. Br Heart J 1955; 17: 552

4. Cha SD, Desai BS, Gooch AS, et al: Diagnosis of severe tricuspid regurgitation. Chest 1982; 82: 726

5. Carpentier A, Deloche A, Hanania G, et al: Surgical management of acquired tricuspid valve disease. J Thorac Cardiovasc Surg 1974; 67: 53

6. Peterffy A, Bjork VO: Surgical treatment of Ebstein's anomaly. Scand J Thorac Cardiovasc Surg 1979; 13: 1

7. Bonow RO, Carabello BA, Kanu C, et al. (2006). "ACC/AHA 2006 guidelines for the management of patients with valvular heart disease: a report of the American College of Cardiology/American Heart Association Task Force on

Practice Guidelines (writing committee to revise the 1998 Guidelines for the Management of Patients With Valvular Heart Disease): developed in collaboration with the Society of Cardiovascular Anesthesiologists: endorsed by the Society for Cardiovascular Angiography and Interventions and the Society of Thoracic Surgeons". *Circulation* **114** (5): e84–231.

8. Deloche A, Guerinon J, Fabiani JN, et al. Anatomical study of rheumatic tricuspid valve diseases: application to the study of various valvuloplasties. Ann Chir Thorac Cardiovasc 1973;12:343–9.

9. Rapaport E: Calculation of valve areas. Eur Heart J 1985; 6 (suppl C): 21.

10. Chapter 66: Valvular Heart Disease. Braunwald's Heart Disease(Elsevier), 9th ed., 1468-1539.

11. Chapters 9: Tricuspid Valve disease. Valvular Heart Disease. Edited by Dalen and Alpert (Little, brown and company). 2th ed., 353-402

12. Chapters 10: Pulmonic Valve disease. Valvular Heart Disease. Edited by Dalen and Alpert (Little, brown and company). 2th ed., 403-438

Abbreviations

A2:	Aortic component of second heart sound
A2C:	Apical two chamber
A4C:	Apical four chamber
A5C:	Apical five chamber
AF:	Atrial Fibrillation
AML:	Anterior mitral leaflet
AP shunt:	Aortopulmonary shunt
AR:	Aortic Regurgitation
ARV:	Atrialized RV
AS:	Aortic Stenosis
ASD:	Atrial septal defect
ASMI:	Anteroseptal myocardial infarction
ATL:	Anterior tricuspid leaflet
AV:	Aortic Valve
AVR:	Aortiv valve replacement
BMV :	Balloon mitral valvotomy
BPV :	Balloon pulmonary valvotomy
BSA:	Body surface area

CCL:	Counterclockwise loop
CABG:	Coronary artery bypass graft surgery
CAD:	Coronary artery disease
CCF:	Congestive cardiac failure
CE:	Cardiac enlargement
CHD:	Congenital heart disease
CHF:	Congestive heart failure
CR1:	
C TGA:	Corrected Transposition of great arteries
CTR:	Cardio thoracic ratio
CXR:	Chest X-ray
D/D:	Differential diagnosis
EC:	Ejection click
ECG:	Electrocardiogram
EDM:	Early diastolic murmur
EF:	Ejection fraction
EMF:	Endomyocardial fibrosis
E/O:	Evidence of
ESD:	End systolic diameter
ESM:	Ejection systolic murmur
ESV:	End systolic volume
ESVI:	End systolic volume index
F/U:	Follow up
FRV:	Functional RV
IAS:	Interatrial septum
ICS:	Intercostal space
IE:	Infective endocarditis

IHJ:	Indian Heart Journal
IRBBB:	Incomplete right bundle branch block
IVC:	Inferior vena cava
IVS:	Inter ventricular septum
JVP:	Jugular venous pulse
LA:	Left atrium
LAE:	Left atrial enlargement
LAD:	Left axis deviation
LAP:	Left atrial pressure
LBB:	Left bundle branch
LBBB:	Left bundle branch block
LCX:	Left circumflex
LICS:	Left intercostal space
LPA:	Left pulmonary artery
LSB:	Left sterna border
LV:	Left ventricle
LVEF:	Left ventricular ejection fraction
LVF:	Left ventricular failure
LVH:	Left ventricular hypertrophy
LVSP:	Left ventricular systolic pressure
LVVO:	Left ventricular volume overload
M1:	Mitral component of S1
Mc:	Mitral closure
MPA:	Main pulmonary artery
MR:	Mitral Regurgitation
MS:	Mitral stenosis
msec:	millisecond
MV:	Mitral Valve

MDM:	Mid diastolic murmur
MVP:	Mitral Valve prolapse
MVR:	Mitral Valve replacement
NYHA:	New York Heart Association
OS:	Opening snap
OA:	Opening amplitude
OTV:	Open tricuspid valvotomy
P2:	Pulmonary component of second heart sound
PA:	Pulmonary artery
PAH:	Pulmonary arterial hypertension
PAO2:	
PAPVC:	Partial pulmonary venous connection
PASP:	Pulmonary artery systolic pressure
PAWP:	Pulmonary artery wedge pressure
PDA:	Patent ductus arteriosus
PFO	Patent foramen ovale
PH:	Pulmonary hypertension
PHT:	Pressure half time
PLAX:	Parasternal long axis
PML:	Posterior mitral leaflet
PND:	Paroxysmal nocturnal dyspnea
PPS:	Peripheral pulmonary stenosis
PS:	Pulmonary stenosis
PSAX:	Parasternal short axis
PSM:	Pansystolic murmur
PSVT:	Paroxysmal supraventricular tachycardia
PTL:	Posterior tricuspid leaflet

PV:	Pulmonary valve
PVR:	Pulmonary vascular resistance
PR:	Pulmonary regurgitation
RA:	Right atrium
RAE:	Right atrial enlargement
RAP:	Right atrial pressure
RBBB:	Right bundle branch block
RCA:	Right coronary artery
RHD:	Rheumatic heart disease
RHF:	Right heart failure
RPA:	Right pulmonary artery
RSOV:	Ruptured sinus of Valsalva
RV:	Right ventricle
RVEF:	Right ventricular ejection fraction
RVEDP:	Right ventricular end diastolic pressure
RVEMF:	Right ventricular endomyocardial fibrosis
RVH:	Right ventricular hypertrophy
RVOT:	Right ventricle outflow tract
RVSP:	Right ventricular systolic pressure
RVVO:	Right ventricular volume overload
Rx:	Treatment
S1:	First heart sound
S2:	Second heart sound
S3:	Third heart sound
S4:	Third heart sound
SCD:	Sudden cardiac death
SLE:	Systemic lupus erythematous
SM:	Systolic murmur
STL:	Septal tricuspid leaflet

SV:	Single ventricle
SVC:	Superior vena cava
SVT:	Supraventricular tachycardia
T1:	Tricuspid component of S1
TA:	Tricuspid atresia
TAPVC:	Total pulmonary venous connection
Tc:	Tricuspid closure
TEE:	Trans esophageal echocardiography
TGA:	Transposition of great arteries
TOF:	Tetralogy Of Fallot
TTE:	Trans thoracic echocardiography
TR:	Tricuspid Regurgitation
TS:	Tricuspid stenosis
TMG:	Tricuspid mean gradient
TRV:	Total RV
TTG:	Tricuspid trans gradient
TV:	Tricuspid valve
TVA:	Tricuspid valve area
TVP:	Tricuspid valve prolapsed
TVR:	Tricuspid Valve replacement
VSD:	Ventricular septal defect
VTI:	Velocity time integral
WPW:	Wolf Parkinson White

About the Author

The author is an interventional cardiologist. He has completed his cardiology training from prestigious G S Seth Medical college and KEM Hospital, Mumbai, India. He has worked as cardiology fellow in Glasgow Royal Infirmary, Glasgow for 1 year. After completion of fellowship he has been practicing cardiology in Gujarat, India for ten years. He has vast experience in management of Rheumatic heart disease (RHD). RHD is still quite common in India and is a major cause of admission in hospitals. The author has rich experience of balloon valvuloplasties. He has already published two handbooks from Authorhouse. Both these handbooks (A handbook of Rheumatic Fever and A Handbook of Aortic valve Disease) are related to this book.

www.ingramcontent.com/pod-product-compliance
Lightning Source LLC
Chambersburg PA
CBHW030839180526
45163CB00004B/1380